BABIES AT THE ZOO

Penguin Chicks

Susan H. Gray

Cherry Lake Press

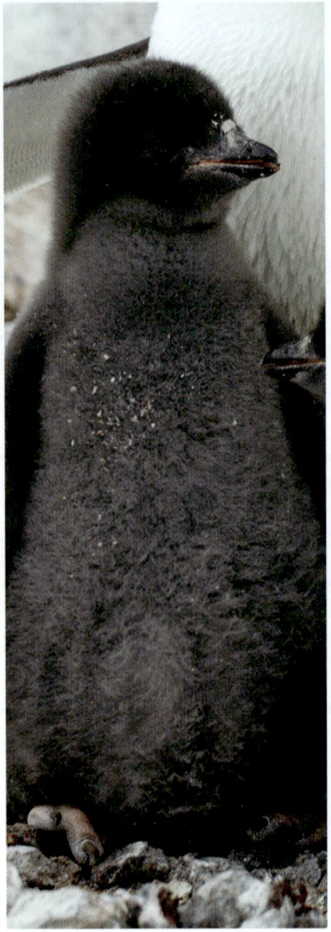

Published in the United States of America by
Cherry Lake Publishing
2395 South Huron Parkway, Suite 200, Ann Arbor, MI 48104
www.cherrylakepublishing.com

Content Advisor: Dominique A. Didier, Professor of Biology, Millersville University
Reading Advisor: Marla Conn, MS, Ed, Literacy specialist, Read-Ability, Inc.

Photo credits: ©Phil West/Shutterstock.com, front cover; ©Tarpan/Shutterstock.com, 1, 2; ©Foto Mous/Shutterstock.com, 4; ©Stefan Wolny/Shutterstock.com, 6; ©By Devon Kubacki/Shutterstock.com, 8; ©Sergey 402/Shutterstock.com, 10; ©Martin Leber/Shutterstock.com, 12; ©Elena Kirey/Shutterstock.com, 14; ©Photodynamic/Shutterstock.com, 16; ©polarman/Shutterstock.com, 18; ©Graeme Snow/Shutterstock.com, 20

Library of Congress Cataloging-in-Publication Data

Names: Gray, Susan Heinrichs, author.
Title: Penguin chicks / written by Susan H. Gray.
Description: Ann Arbor, Michigan : Cherry Lake Publishing, 2020. | Series: Babies at the zoo | Includes index. | Audience: Grades K-1.
Summary: "Read about cute penguin chicks and why zookeepers might put bandages on unhatched eggs. This level 3 guided reader book includes intriguing facts and adorable photos. Students will develop word recognition and reading skills while learning about how these baby animals learn and grow, what they eat, and how they socialize with each other. Book includes table of contents, glossary, index, author biographies, sidebars, and word list for home and school connection"—Provided by publisher.
Identifiers: LCCN 2019034163 (print) | LCCN 2019034164 (ebook) | ISBN 9781534158986 (hardcover) | ISBN 9781534161283 (paperback) | ISBN 9781534160132 (pdf) | ISBN 9781534162433 (ebook)
Subjects: LCSH: Penguins—Infancy—Juvenile literature. | Zoo animals—Infancy—Juvenile literature.
Classification: LCC QL696.S473 G73 2020 (print) | LCC QL696.S473 (ebook) | DDC 598.4713/92—dc23
LC record available at https://lccn.loc.gov/2019034163
LC ebook record available at https://lccn.loc.gov/2019034164

Cherry Lake Publishing would like to acknowledge the work of the Partnership for 21st Century Learning, a Network of Battelle for Kids. Please visit http://www.battelleforkids.org/networks/p21 for more information.

Printed in the United States of America
Corporate Graphics

Table of Contents

About the Author

Susan H. Gray has a master's degree in zoology. She has written more than 150 reference books for children and especially loves writing about animals. Susan lives in Cabot, Arkansas, with her husband, Michael, and many pets.

Pipping Penguins

Penguins are large birds. Their backs are black or dark blue. Their bellies are white.

Penguins live in the southern half of the world. Some live in zoos.

All penguins have wings. But they cannot fly. They use their wings to swim. Penguins spend lots of time in the water. They eat fish and other sea animals.

Can you see the hole the chick is making in the egg?

Most penguins lay one or two eggs. A baby **chick** is inside each one. After 5 to 10 weeks, that chick wants out!

The chick first pecks a hole in the shell. This is called pipping. The baby keeps working. It can take 3 days to **hatch**!

How can a zookeeper save this unhatched baby chick?

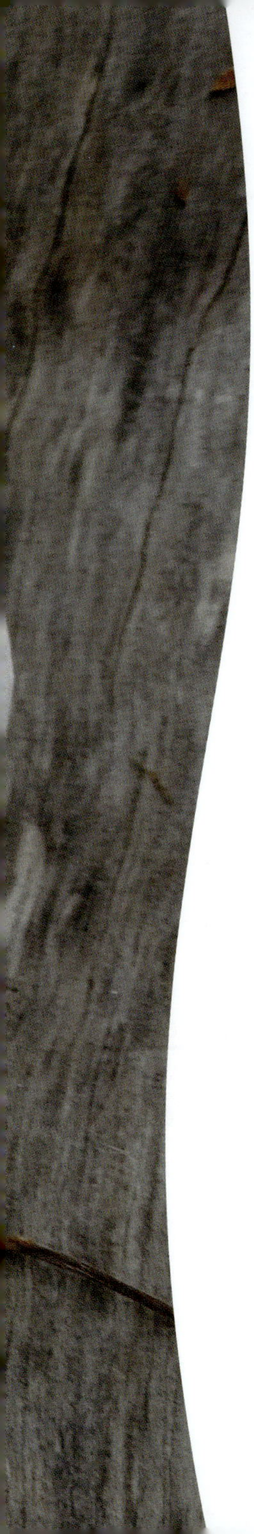

Taking Care of Penguins

Zookeepers keep an eye on eggs. Sometimes, one cracks by **accident**. So zookeepers put a special **bandage** on it. This is so it will hatch okay later.

Zookeepers give fish and **vitamins** to the birds. They also teach penguins to stand on a **scale**. Then zookeepers can weigh them.

Parents and Chicks

Baby penguins are not black and white. They are gray or brown. Their feathers are soft and fluffy.

Penguins live in big groups. Sometimes a baby gets lost. But parents can find it. They know their baby's voice!

A penguin stays with its parents for up to a year. Then it lays its own eggs.

Soon, the pipping begins!

Find Out More

BOOK

Salomon, David. *Penguins!* New York, NY: Random House, 2017.

WEBSITE

San Diego Zoo—African Penguin
https://kids.sandiegozoo.org/animals/african-penguin
Find more penguin facts here and click on a link to a live "penguin cam."

Glossary

accident (AK-sih-duhnt) something that was not expected to happen

bandage (BAN-dij) something used to cover a cut or sore

barbs (BARBZ) sharp points that stick out

chick (CHIK) a very young bird

hatch (HACH) when an egg breaks open and a baby bird comes out of it

scale (SKAYL) a device used for weighing things

vitamins (VYE-tuh-minz) things added to food that all animals need for a healthy life

zookeepers (ZOO-kee-purz) people who take care of zoo animals

Home and School Connection

Use this list of words from the book to help your child become a better reader. Word games and writing activities can help beginning readers reinforce literacy skills.

a
accident
after
all
also
an
and
animals
are
baby
backs
bandage
begins
bellies
big
birds
black
blue
brown
but
by
called
can
cannot
care
chick
chicks
cracks
dark
days
each
eat
egg
eggs
eye
feathers
find
first
fish
fluffy
fly
for
gets
give
gray
groups
half
hatch
have
hole
how
in
inside
is
it
its
keep
keeps
know
large
later
lay
lays
live
lost
lots
making
most
not
of
okay
on
one
or
other
out
own
parents
pecks
penguin
penguins
pipping
put
save
scale
sea
see
shell
so
soft
some
sometimes
soon
southern
special
spend
stand
stays
swim
take
taking
teach
that
the
their
them
then
they
this
time
to
two
unhatched
up
use
vitamins
voice
wants
water
weeks
weigh
white
will
wings
with
working
world
year
you
zookeeper
zookeepers
zoos

Fast Facts

Habitat: Coastal areas

Range: Southern Hemisphere

Average Height: Little blue penguin, 14 inches (36 centimeters); emperor penguin, 46 inches (117 centimeters)

Average Weight: Little blue penguin, 3 pounds (1.5 kilograms); emperor penguin, 80 pounds (36 kilograms)

Life Span: 6 to 20 years

Anatomy: Penguin tongues have small, backward-pointing **barbs**. These help them hold on to slippery food.

Behavior: Penguins can hold their breath underwater for up to 15 minutes.

Index